RICHARD MABEY is one of our greatest nature writers. He is the author of some thirty books including the bestselling plant bible, *Flora Britannica*, *Food for Free*, *Weeds*, *The Cabaret of Plants* (both Profile) and *Nature Cure* which was shortlisted for the Whitbread, Ondaatje and Ackerley Awards. His biography *Gilbert White* (Profile) won the Whitbread biography Award. He is a Fellow of the Royal Society of Literature, and Patron of the John Clare Society, and lives in Norfolk.

TURNED OUT
NICE AGAIN

TURNED OUT NICE AGAIN

On Living With the Weather

RICHARD MABEY

PROFILE BOOKS

This paperback edition published in 2019

First published in Great Britain in 2013 by
PROFILE BOOKS LTD
3 Holford Yard
Bevin Way
London WC1X 9HD
www.profilebooks.com

1 3 5 7 9 10 8 6 4 2

Typeset in Fournier by MacGuru Ltd

Printed and bound in Great Britain
CPI Group (UK) Ltd, Croydon CR0 4YY

A CIP catalogue record for this book is available from the
British Library.

ISBN 978 1 78125 181 2
eISBN 978 1 84765 895 1

CONTENTS

1. Turned Out Nice Again   1
2. Air-songs and Moon-bows   17
3. Black Dog   35
4. Halcyon Days   53
5. Mixing Memory and Desire   71
6. The Storm Clouds of the
   Twenty-first Century   87

For Tim, in all weathers

1

TURNED OUT
NICE AGAIN

WHEN HURRICANE SANDY'S SIEGE of New York in the autumn of 2012 was smartly followed by waves of marauding floods in Britain, the closing acts in a year of numbing gloom and damp, the idea of 'global warming' began to sound a rather black joke. Ten years ago, some optimists were relishing the prospects of olive groves on the South Downs. Now it looks as if we may be heading full steam for the state of Newfoundland.

But scientists, if we'd listened properly, have always insisted that climate change can't be neatly translated into weather patterns. It's likely to generate incoherence, extreme events. Climate may be the big slow-moving backdrop, but weather is what happens here and now, to our settlements and landscapes, to *us*. In that sense, it's part of our popular culture. And that is what I will be exploring in this book, how weather enters and affects our daily lives in Britain, how we talk and write about it, make it the stuff of nostalgia and dreads and, in these uncertain times, how it changes the way we think and feel, about ourselves and the future.

Let me give you an example of what I mean by something that happened to me back in the 1980s – as turbulent a time as today, despite our selective memories insisting otherwise. It was an autumn afternoon, and I was meandering through a favourite wood in the Chilterns full of ancient, cranky beech trees. Frithsden has always been an epic weather theatre, a place where freak frosts can scorch the bracken as early as September, and south-westerly gales routinely strew the ground with 300-year-old gothic pollards. It was becoming a kind of woodwreck by then, I suppose, but also gave off the aura of a wood-henge; and whatever melancholy I felt walking among the fallen was always balanced by a frisson of excitement that something wonderfully Promethean was happening inside the green chaos.

Well, on that particular afternoon the weather upped the stakes. Out of a clear blue sky (how we love our weather metaphors!) it began to pour, in sheets. The rain was ferocious, spattering off the golden leaves in silver jets. The whole wood began to change colour, the trunks slicking to slate grey, next year's beech-buds glistening like glazed fruit. I huddled under the nearest holly and realised that I'd gone to ground right next to the remains of a dear departed. It was the tree I called the 'Praying Beech', on account of two branch stubs that had fused across it just like a pair of clasped hands. Four years earlier it had been split open by a lightning strike. Bees had nested in the hollow gash. Then it was toppled in a storm. Now this gargantuan supplicant, half as tall as our parish church, was prostrate on the ground. And it was liquefying in front of my eyes. The rain was hammering drills of water at the already rotting trunk, and flakes of bark, fungal ooze, barbecued dregs from the lightning-charred

heartwood, began to drip onto the wood-
land floor like thick arboreal soup.

Peering out from my bush I was mes-
merised. I was witnessing the dissolution
of a tree, but also what felt like the beginning of something new, the elements of
forest life returning to the crucible. The
alchemy wrought by that storm changed
my whole view of weather and the resil-
ience of nature.

By any standards it was a spectacular
weather event. If I hadn't been the only
witness, it could have become a star piece
of local mythology, part of that ceaseless,
nagging narrative we British have about the
weather. The poet Samuel Coleridge, one
of the greatest writers on what in his day
were called 'Meteors', would have relished
the bizarre vision of a dissolving tree. On
26 July, 1802, when a day of topsy-turvy
Cumbrian weather had left the sky dotted
with flotillas of motionless clouds, looking,
he thought, 'like the surface of the moon
seen thro' a telescope', he'd had a brain-
wave. Why didn't he write a set of posters
– 'Playbills' he called them – 'announcing

each day the performance by his supreme Majesty's Servants, the Clouds, Waters, Sun , Moon, Stars.'

He never got round to it, but I reckon his scheme might go down well today. The playbills would be rather eye-catching, stuck on parish noticeboards alongside the programmes of the local dramatic society. 'Melting tree on the common!' 'Lightning scar on church door!' 'Five-foot icicles hanging round the council offices – keep a careful distance!'

We're often mocked for our national obsession with weather, and the fact that some blindingly obvious remark about it is often the first greeting we make to a fellow human. 'Turned out nice again' we say, or 'The winds got up'. Our comments are usually banal catch-phrases, hardly con-versation at all, signs perhaps of our stiff – maybe *frozen*-stiff – upper-lips. But I find it heartening that we use these coded phrases as a kind of acknowledgement that we're all in the weather together. *Of course* we should be preoccupied. It's the one circum-stance of life which we share in common. It affects our bodies, our moods, our behaviour, the structure of our environments. It can change the cost of living and the likelihood of death. It is a kind of common language itself.

And though much of the time we complain about our climatic lot, about our seemingly inexorable legacy of insidious rain and grey skies, there's a little bit of us that relishes rough weather, just so long as it doesn't move into truly malevolent mode. So, we swing between sulky resentment and playful derring-do. The municipal gritters never arrive on time, our plumbing is a disaster, but come the first decent snowfall and we are out playing truant with the toboggans. On the first day of December our local pub in Norfolk throws down the gauntlet by announcing – in its own version of the Coleridgean playbill – a 'Guess the Date of the First Snowfall' competition, pinned on a board next to the biggest wood-burner in the district. We can make any pastime into a winter sport. I once watched the annual Oxford and Cambridge rugby match played with mad gallantry in four inches of snow and a temperature of minus six C. Glastonbury is now as much a mud festival as a music festival, and has turned the Wellington boot into a fashion item.

Our creative sparring with the climate flourishes when it comes to clothing. The 'Country Life' style has always been ridiculed for its pomposity and obvious discomfort, but vernacular weather togs are another matter. I love seeing folded newspaper sun-hats and knotted handkerchief on the beach and fieldworkers donning fertiliser bags in summer storms. Once, by the River Dove in Derbyshire, I watched a gang of five-year-olds picking the rhubarb-sized leaves of butterbur to make themselves umbrellas during a downpour. I've no idea whether they were just aping modern brollies, or had the same instinctive sympathy with plants that led the ancient Greeks to name butterbur *petasos*, meaning a broad-brimmed hat. On another cold and showery day in 1802, Dorothy Wordsworth and her beloved brother William buttoned themselves up together

9

in a big 'Guard's coat', and Dorothy coyly confessed that she 'liked the hills and the rain the better for bringing us so close to one another …'

These ambivalent, not so say contrary, responses to the weather – fury at the 'wrong sort of snow on the line' co-existing with a breakfast-table thrill at hoar-frost turning the trees to lacework – are special to Britain. They happen because we haven't really the foggiest idea about what, day on day, to expect, so that any slightly untoward disturbance of the atmosphere is regarded as an unnatural affront or, then again, an unexpected benediction. Sudden snow-falls and un-forecast heat-waves throw us equally. Because of where we live, on an island in the middle of the Atlantic Storm Belt, just offshore from a huge, breathing, land-mass, our meteorological lot is messy and erratic, whether we like it or not. We can't acclimatise, reconcile ourselves to these repeated bolts from the blue. In reality our climate is quite mellow. We don't have to live with active volcanoes or sudden tsu-namis. The temperature has only exceeded

100 degrees three times in the last hundred
years. The heaviest rainfall in a single day
was eleven inches in Martinstown Dorset on
18 July, 1955. When you compare that with
the several feet that can fall in a couple of
hours in a tropical monsoon you can get our
weather in some kind of perspective. What
we really suffer from is a *whimsical* climate,
and that can be tougher to cope with than
knowing for sure you're going to be under
three feet of snow every December.

And hanging over all of us now is
that more sinister unpredictable, climate
change. It's already happening, and there
are few encouraging signs that we're willing, or able, to do anything about it. But
how it might translate into local weather is
hard to predict. Even harder is imagining
how we and the rest of creation will react.
Living organisms aren't passive victims
even when the climate they're experiencing
is changing at unprecedented speed.

But if we are bequeathed a new climate,
of whatever sort, its bouts of maverick
weather won't necessarily be unfamiliar.
The best antidote to an attack of 'we've

never had it so bad' is simply to look back clearly at the past. Every extreme and nuance of weather has been experienced in Britain before, at least for a spell. And for at least 400 years, writers, painters, scientists and folk in the street have left records of our legacy of outrageous, beautiful, violent, glorious, mysterious and simply down-home-ordinary weather, and how they – and we – reacted to it. We have more weather proverbs than the Inuit, proverbially, have descriptions for snow. Constable and Turner's paintings brim over with weather, as do the works of modern artists like Kurt Jackson. We have weather symphonies and weather nursery rhymes. And it would be hard to find an English writer whose diaries don't carry, under the dominant melody of their daily lives, the choral hum of sun-dried grass and windblown leaves and subterranean water. Dorothy Wordsworth was Coleridge's gentler, domestic soul-mate, and in March 1802 recalls watching 'Little Peggy Simpson standing at her door catching the hail stones in her hand'. Gerard

Manley Hopkins, the great synaesthetist, perceived sheets of bluebells as floods, and clouds as solid rocks. On 19 June, 1848 he saw 'two beautiful anvil clouds so low on the earthline in opposite quarters, so that I stood between them.' For about thirty years I kept a kind of nature diary myself but it was constantly interrupted by sulky notes about my internal weather – a reminder that the climate is also an influence on and a metaphor for our well-being. Thomas Hardy, most Martian-eyed of diarists, wrote one freezing January that 'Cold weather brings out upon the faces of people the written marks of their habits, vices, passions, and memories, as warmth brings out on paper a writing in sympathetic ink.'

But for me the master journalist, and a figure whose unsentimental but always luminously acute notes on the weather will recur throughout this book is the eighteenth century curate Gilbert White, of Selborne. White was the first truly literary naturalist and the journal he kept for the best part of forty years, raised weather writing to a new level. If most diarists are

13

essentially weather-scapists, White was a weather story-teller. His journals are full of spare, glittering miniatures that often have the depth and rhythm of haiku. 31 March, 1768: 'Black weather, Cucumber fruit swells, Rooks sit' – the ambivalent progress of March caught in a three-act drama of seven words. Or this entry from 1 February, 1785 that could be a scene from a symbolist film: 'On this cold day about noon a bat was flying round Gracious street-pond, & dipping down & sipping water, like swallows, as it flew: and all the while the wind was very sharp and & the boys were standing on the ice!'

The skill of all these diarists is that they catch the feel of weather through ordinary domestic detail. White knew it was really cold when the pisspot froze under his bed, just as I did – before we had central heating – when there was frostwork on the *inside* of the windows. His heat-wave threshold was when the meat went off in the safe; mine was when there were tab-bubbles on the roads. Samuel Pepys caught the extraordinariness of a scorching July day in 1667 by describing how he slept, daringly and for the first time since he was a boy, with only a rug and sheet upon him.

This intimacy of detail has now been adopted by the weather forecasters. Their invention of 'spicks and spots of rain' has entered the vernacular. In 2012 the term 'grass frost' became fashionable, a friendly phrase that describes exactly what you could see out of the window first thing without having to leave the house and thrust a fork in the ground. Weather forecasters have become our new shamans, and the forecast has, in a sense, become *part* of the weather, an affecting, emotional experience as well as a detached prediction.

The forecast reminds us, crucially, that the weather, in our culture and our psychology, is intricately linked with *time*, and especially with time's familiars, memory and expectation. Sometimes weather leaves physical relics of its fulminations. Gilbert White wrote an electrifying account of the tangled banks of Hampshire's deep hollow lanes, and how they had been formed by centuries of floods and frosts. Weather had been preserved in the aspic of geology.

So in the following chapters I'm going to look at this constant thrum in our lives, which stretches from global jet-streams to the pulses of our individual cells. At how we celebrate the good times, cope with the bad, worry about the unknown; and at our anciently hopeless attempts to reconcile our dreams of a calm life under the skies with the elements' gratuitous rampagings.

2

AIR-SONGS AND
MOON-BOWS

AT THE END OF HIS BOOK *Wildlife in a Southern County*, the nineteenth century nature writer Richard Jefferies adds a note about what he calls 'Noises in the Air'. He describes how Wiltshire haymakers often reported hearing mysterious, distant booms when the weather was exceptionally still and calm. They believed something – possibly something supernatural – was happening 'in the air'. But Jefferies, down to earth as ever, was convinced it was the sound of naval guns, on exercises in the Channel over fifty miles away. He could have been right. Low-frequency sounds can travel long distances when the air is warm and still, become aural mirages, analogous to the optical illusions that occur in just the same conditions.

Jefferies was hyper-sensitive to the

nuances and exceptionalities of weather sounds. He boasted he could distinguish species of trees by the particular rustle their branches made in the wind, as could Giles Winterbourne in Thomas Hardy's *The Woodlanders*. In July 1874, Jefferies records another curious and unexplained aerial noise, an electric crackle passing through a wheatfield, increasing as the wind blew. It probably wasn't mysterious at all, just the noise of ripe wheat grains being popped out by their seedheads. But outdoor noises of inexplicable origin can be as disconcerting as night-time creaks inside an old house. I once heard what sounded like a battery of cap-guns going off as I was walking across a heath on a blazing June day. I thought it might be a gathering of anxious stonechats, familiars of gorse bushes, whose 'tchak' alarm calls have just this crisp, percussive quality. Then something hit me smartly in the face, and I realized the sound was hundreds of gorse-pods exploding in the heat and hurling their seeds into the air. Maybe this was what has happening in Jefferies's wheatfield.

We love to tell tales about strange weather occurrences. And we're oddly proud of them. They didn't just happen, they happened to *us*. There are plaques on seaside buildings to commemorate the high watermarks of historic floods. National forecasters give proper attention to record-breaking weather, like the fact that 2012 was the wettest year in England since, as they say 'records began'. But they also announce, as if they are giving away end-of-term prizes, the highest temperatures, the deepest snowfalls, the earliest frosts, at a scale that would be more appropriate for a local history project. 'Foggiest Day on Tyne for ten years' is hardly of record-breaking interest for the rest of us, but it is for the people who live there. It's their weather. There is, from our ringside seats at the oddball-weather circus, an intriguing interplay between freak-show and something more distinctively local and neighbourly. On 8 April, 1979, eleven amateur footballers in Gwent were each struck by lightning as they ran off the field during a thunderstorm. None were badly hurt. During Britain's fiercest tornado, on 21 May, 1950, a cat was seen in full flight through the air in Leighton Buzzard, all four legs akimbo in an automatic balancing act. In Hemel Hempstead, at the height of the July 1983 heatwave, the municipal dahlia beds caught fire when someone dropped a cigarette onto earth that had turned into a baking cake of inflammable fertiliser. And on 13 February, 1879, the Reverend Francis Kilvert baptised an infant in his little church in Clyro, Radnorshire, 'in ice which was broken and swimming about in the font'.

You could list such oddities endlessly – ice-meteors, hayfield twisters, rains of herrings. I've seen a few myself. I've been in one of the 'red-rains' that happen when southerly winds blow immense clouds of Saharan dust north, so that they stain cars and washing and even, on one occasion, the open pages of a prayer-book during a burial. I remember being thrilled at finding this exotic exhalation of Africa – a kind of climatic spice – in our Home Counties backyard; that was until I learned of another Saharan dust-blow of barely credible proportions. Every winter forty million tonnes of desert sand is sucked up from Chad by the wind and whisked 5,000 kilometres across the Atlantic to gently descend on the Amazon basin. It is the main source of new minerals in Amazonia, a fertilisation on a global, Gaian scale without which the rainforest couldn't survive.

I've also, just once, witnessed a glazed frost, a phenomenon of scary beauty which occurs when rain falls onto a landscape locked under air at sub-zero temperatures and freezes on impact, so that all solid objects rapidly appear to have become crystallised in glass. I noted in my diary for that day in January 1979, that 'as raindrops hit the window panes they made distinctive harmonic rings, and soon built up a quarter of an inch of glaze.'

But again home-team pride is trumped by awesome encounters in the weather's Premier League. In Britain's greatest twentieth-century ice-storm on 27 January, 1940, cats were iced to branches and birds killed in flight as their wings froze solid.

Telegraph wires rotated under their pay-load of ice (one stretch in Gloucestershire carried eleven and a quarter tons between just two posts) so that they were adorned by upward-pointing icicles. Gilbert White – and what an extraordinary record of natural wonders he kept – described a miniature and picturesque ice-storm that might have been specially designed as a *divertissement* for his gentleman's landscape garden. It happened on 10 December, 1784, and he had never seen anything like it before.

> Being bright sunshine, the air was full of icy spiculae, floating in all directions, like atoms in a sun-beam let into a dark room. We thought them at first particles of the rime falling from my tall hedges; but were soon convinced to the contrary, by making our observations in open places where no rime could reach us. Were they watery particles of the air which froze as they floated; or were they evaporations from the snow frozen as they mounted?

24

I love this image of White, enchanted by the glittering tinsel in his outlet, but not so dazzled that the amateur scientist in him didn't bustle out for an alfresco experiment in the adjoining field.

Bizarre weather, from teasing crackles in the air to falling ice-bombs can, in principle, happen anywhere. But reading about it in these vivid diary records, stretching over the centuries, it doesn't seem at all random. It isn't just that each event is, so to speak, date-lined, occurring at a particular moment in a specific place, but that they seem to fit those niches almost ecologically. A soccer team struck by lightning, all together on the exposed pitch. The thin plates of mist that rise over sandy heaths on August evenings – still for me the most atmospheric evocation of the change from summer to autumn – which come up no higher than your chest, so that you can gaze down on them as if you are in an airliner, or playing at being Gulliver. And Coleridge again, and his breathtaking, intensely-located vision of what he called a *smoke-flame* in August 1800, a pillar of fire-coloured clouds soaring up through the crevasses round Derwentwater. They all feel so *right*, as properly placed as a hatching chrysalis on a grass-stalk, or a swallow arriving back in its barn. (And other organisms can be more than metaphors in these small dramas. One cold and windblown day in June I found a troop of house-martins strafing for insects deep inside the shelter of a canal lock, a local refuge inside the local weather.) These fine-tunings between weather and habitat – you might call them weather accents or dialects – seem to me the meteorological equivalents of biodiversity, a tribute to the variety and eccentricity of Britain's landforms.

And this is surely the reason we find the shipping forecast so evocative, even though most of us are hazy about where the great stretches of ocean it describes are situated. Yet their names – brooding, wind-tossed, pewter-grey names – seem to be emanations of the sea-parishes themselves: Lundy, Fastnet, South Utsire, North Utsire. They're the music of weather's local distinctiveness. Sean Street's poem 'Shipping Forecast, Donegal', catches their sense of being *incantations* – not just respectful tributes to sea and weather, but call-signs from the home-patch:

Fisher, German Bight, Tyne, Dogger
This pattern of names on the sea –
Weather's unlistening geography...
this minimal chanting,
this ritual pared to the bone
becomes the cold poetry of
    information.

The exceptionality of local weather can produce moments and places that transcend the physical bluster of freak tornados and glazed frosts. They can be dramas of pure sound or light, so that to be present is like walking into an art installation. Coleridge – who always seemed to be in the right place – once witnessed a moonbow, the spectral arc of colour formed when moonlight passes through fine rain or water spray. He was in Cumbria on 22 October, 1801. 'Thursday evening, half past 6. All the mountains black and tremendously obscure … At this time I saw one after the other, nearly in the same place, two perfect Moon Rainbows … It was a grey, moon-light mist-colour.'

I've never seen a moon rainbow, but I have seen a cave rainbow, which is the next best thing, and even more topographically specific. It was 1986, and I was sixty miles east of Coleridge's Lakeland retreats, making a film about the limestone country of the Yorkshire Dales. We had decided to film a sequence down in the aptly named Weathercote Cave, partly because it's one of the most extraordinary chambers carved out by the watercourses that catacomb the Dales, and partly because Turner had painted weatherscapes *inside* the cave. Weathercote has a waterfall tumbling into it, lit up by the sun where part of the cave roof has collapsed, so that it shimmers with iridescence. It's no wonder Turner was entranced by it during his painting tours of the Dales. He first visited it in 1808, and made the perilous seventy foot descent just to see the marvel of a cascade of water falling out of the sun. But on his second trip eight years later, he couldn't even get inside. The rain had been so heavy for weeks that the underground rivers were in spate and bursting like fountains through the hillsides. Weathercote itself was half-full of water, and Turner had to make do with a rapid drawing made from the top, noting in his sketch-book 'Entrance Impossible'. He marked the time of day as mid-afternoon, with the sun shining from the left, and low enough to throw a beam through the boiling spray of the waterfall. The result, an ectoplasmic 'curve of prismatic colour', hovering over the sunless depths below, is clearly visible in his finished watercolour of 1818.

The tumbling waterfall still fills the cavern with an eerie and almost luminous mist and I saw rainbows in the spray, too, when I climbed down. It was cold and slippery inside and I edged onto a fallen block of limestone known as Mahomet's Coffin, which hangs suspended between cave roof and floor. Momentarily, to keep my balance, I leaned forward slightly, and abruptly one of the rainbows flipped over on its side and formed a circle, completely surrounding me at chest level, like a fallen halo. High above me, joining sky and earth in another way, flycatchers were swooping down into the cave and hawking for midges in the sunbeams.

Four years earlier I'd made a trip to the western reaches of England, after another weather phantom. I'd read about a wood on the River Fal in Cornwall that was tidal at the spring equinox, a unique mix of tree and wave; and that if the wind was a stiff sou'wester that day, the water would rise high enough into the wood for you to have the surreal vision of primroses flowering under the sea. There was one other thing. The Fal flows through the china-clay beds below Bodmin. Much of the clay has been mined out lately, but when it reaches the estuary the river can still be as milky as whey.

So, on the afternoon of 21 March, first day of spring, I perch under the oaks in Lamorran Wood and wait for the equinoctial high tide. There are piping curlews overhead, and a thin rime of salt on the lowest branches, maybe a relic of earlier inundations. When the high water seeps up to where I'm sitting, it's not quite the dramatic mix of wood and water I'd hoped for. It laps milkily and rather sedately around the primroses and golden saxifrage, but doesn't *flood* the wood as I'd dreamed it might.

But in the night, out of my sight, it did. A south-westerly gale had blown up and the spray was lashing the second-storey windows in my hotel. Next morning Lamorran Wood and the whole Fal estuary was a scene of devastation, littered with flotsam, and with a thin veil of white clay covering the whole of the low-lying land.

The theatre of the weather includes music, too, beyond the bounds of Jefferies' 'noises in the air'. Flora Thompson, working as a post office telegraph operator years before she wrote *Lark Rise to Candleford*, loved to listen to the wind singing in the telegraph wires, *her* wires; and in their busy metallic hum she liked to imagine their role as a 'golden highway for ... messages to traverse from friend to absent friend'. I have the same kind of feeling for the rattle a sea-wind makes in the rigging of moored dinghies. Extraordinarily, this *vox loci* has, as yet, no popular name, yet for me hearing it at night on the Norfolk coast sets my skin tingling; its siren's song about the lure of the water's edge as powerful as the urgent calls of wading birds going to roost.

This is accidental wind music, as are the deep bassoons a few hollow trees become in gales. But there is deliberate weather music, and literal wind instruments, too. The Aeolian harp is the best known, named after Aeolus, the mythological keeper of the winds, and popular from the sixteenth to nineteenth century. It consisted of a wooden box about three feet long, fitted with gut strings of different thicknesses, which was placed on a window-sill or any outside ledge. The strings were tuned in unison, and the vibrations the wind produced in them varied according to the strings' thickness and generated ethereal harmonics and chord-like effects.

Much larger versions are now made as

garden ornaments or outdoor installations. But the biggest Aeolian instrument never got beyond the drawing board. In 1980, the architect H. T. Cadbury Brown put forward a proposal for a memorial to the composer Benjamin Britten, who had lived at Aldeburgh, and whose music was so expressive of the temperamental weather of the Suffolk coast. Brown's musical obelisk would be like a giant oboe, a wooden column erected on Aldeburgh beach, and drilled with holes which would whistle in the wind. When a gale off the sea built up enough ferocity the column would produce the two notes used by the crowd in Britten's opera *Peter Grimes*, when they call out obsessively for the mad, disaffected fisherman.

The idea was never taken up, but the artist Maggi Hambling has created her own haunting memorial to Britten on the Aldeburgh shingle. *Scallop* is a giant steel shell, facing out over the North Sea. It's mesmerising seen from a distance, shape-shifting from mollusc to fairy-tale sailing-ship to seaweed forest; but close-to it makes you turn your head, to listen to the reflected roar of the sea – which on this coast, has drowned not just fictional characters like Peter Grimes but whole settlements. Half of the medieval town of Dunwich, six miles up the Suffolk coast, lies under the sea, and has its own legendary weather music, the submarine bell-peals of drowned churches.

There is often a dark shadow behind the most seductively dramatic and beautiful weather. On the top rim of the scallop, Hambling has drilled out a calligraph which spells out a line from *Peter Grimes:* 'I hear those voices that will not be drowned.'

3

BLACK DOG

IN EAST ANGLIA where I live, everyone knows one particular skin-crawling weather legend. On Sunday 4 August, 1577, the market town of Bungay in Suffolk was visited by what witnesses called a 'straunge and terrible Wunder'. The church was struck by a violent electric storm, with such thunder and lightning 'as was never seen the lyke'; and in the same instant, 'a horrible shaped thing' passed down the aisle, causing nightmarish casualties. Two men, kneeling at prayer, had their necks jerked backwards and broken. Another, according to a local chronicler Abraham Fleming, received 'such a gripe on the back that therwithall he was presently drawen up togither and shrunk up, as it were a peece of leather scorched in a hot fire: or as the mouth of a purse or bag drawen together with a string.'

The phenomenon was almost certainly an instance of the rarest and most eerie of weather events, ball lightning. No convincing explanation has been found for this apparition, which is usually associated with storms, and appears as a bright, vaguely spherical ball of electromagnetic energy, capable of moving through the air, breaking windows and entering buildings, crawling up walls and along floors, and occasionally killing people stone dead.

But the parishioners of Bungay saw something extra that fateful Sunday. Not just bright 'flashes of fire', but a 'dark companion', a great black dog, which tore marks out of the church door with its talons. They're still visible if you're impressionable enough, as they are on the north door of Blythburgh church, visited by the diabolical beast on the same day.

The Great Black Dog, often known as Black Shuck, has since become the most famous creature of East Anglian folklore. He's been spotted across the region for the last four centuries, and a sighting is usually supposed to presage death or serious trouble, though this isn't borne out by the subsequent experiences of the witnesses.

Black creatures are widespread symbols and portents of devilish states of affairs, and it's not surprising that two centuries later the phrase 'black dog' became a metaphor for another kind of affliction – melancholy or depression. Scientifically curious observers noticed that this dark invader of the spirit often appeared alongside severe weather, just as, allegorically, Shuck had manifested himself during the Bungay tempest. Samuel Johnson was the first writer to use the term in print. Writing to his friend Mrs Thrale in 1783 he says 'When I rise my breakfast is solitary, the back dog waits to share it, from breakfast to dinner he continues barking …' Johnson's biographer Boswell also suffered, his dour temper always markedly worse in foul winter weather. By the time Winston Churchill had popularised the phrase for his own bleak moods during World War Two, medical science was becoming aware that there were clear links between weather and human psychology, though no precise conditions, like SAD, had yet been defined.

I suppose I am what is usually described as 'weather sensitive'. I'm slow to adapt to cold, get freaked by incessant wind, become morose and torpid in the dark winter months. What's mystifying to me is that *everybody* isn't a registered member of this club. Perhaps, in an offhand way, we are, since 'feeling under the weather' is the most commonly-used metaphor for being off-colour. If you do no more than track across the exterior of our bodies you'll realize we are a landscape of tissue at the total mercy of the elements. Sunshine can give us burns, sunstroke, melanomas, prickly heat, photosensitive rashes, even blindness. Persistent wind can bring on dehydration, wrinkles, maddening tics. Cold can conjure up frostbite, chilblains, hypothermia, a dangerous lowering of the pain threshold.

When you go below the body's surface, an even more bewildering array of other sensitivities emerges, chiefly due to the body's response to air-pressure. Our insides, from individual cells to whole digestive systems, are labyrinths of gaseous cavities and bags of fluid, so it's no wonder that the genetically susceptible react dramatically to the rapid changes in outside pressure that accompany the passage of a weather front, especially a low. The volume of the fluids in joints expands, aggravating rheumatic complaints. The width of blood vessels and the capacity of the lungs change. Retinas can detach and the business of giving birth speeds up. During the passage of an extreme low front, hospital admissions for problems as diverse as schizophrenia and phantom limb pain more than double. The increase in strokes is so marked that in Germany there is a 'Metalert' to warn doctors of approaching pressure troughs.

It's these low fronts, and the dinge that so often accompanies them, that always seem to give me trouble, bringing on symptoms that are certainly doggish and black – anxiety attacks, irritability, a heaviness of mind and body that teeters on the edge of outright depression. They're worse in the winter, when the decline of daylight, acting through the pineal gland, reduces the levels of hormones like melatonin and serotonin to levels close to those that occur in sleep – or hibernation. In susceptible individuals – maybe five to ten per cent of the population – this manifests itself as full blown Seasonal Affective Disorder, whose serendipitous acronym, SAD, says it all.

But it has never been as straightforward as that for me. When I first started a nature diary I began to find notes on my spells of weather malaise were sneaking in among records of the arrival of summer migrants and the autumn leaf-colour change. It didn't take me long to spot that the symptom clusters often appeared at the same time, sometimes even the same days each year, like those annually recurrent spells of weather known as 'Buchan's Periods'. I began to feel like a seasonally driven-organism myself. If, as I often persuaded myself, I'd got the winter blues, then I'd often got the April, midsummer and autumnal blues, too.

When this all became entangled with a bout of real depression in my middle age, I discussed it with my psychiatrist, a lean Scot with an uncompromisingly direct approach.

He was dismissive of my self-diagnosis: 'It's not your biochemistry that's off, laddie. You just don't like what's happening out there.' And he was partly right. When I'm straight with myself, I can see my bouts of weather unease have a lot to do with the crushing of expectations. I know the kind of events that act as triggers now. I find bumble-bees frozen to the crocuses on the first day of spring. The swifts fail to arrive on time, blocked themselves by low pressure over the continent, and Ted Hughes's famous cheer from the terraces, 'They're back, which means the globe's still working …' – becomes an anxious cry in the wind. The re-enactment of seasonal weather events, of that proper order of things that anchors us not just in the present moment but in the long rhythm of our lives, breaks down too often for comfort. Who wouldn't get depressed! Seasonal affective disorders may be biochemical in part, but they are also cognitive. They're about our interpretations of 'what's happening out there', about the tarnishing of childhood memories, about dashed hopes and lost moorings.

But our memories are fallible, our belief that there is a proper order of things hugely over-simplified, and our interpretation of what we are experiencing often emotionally warped and highly selective.

Nothing demonstrates the subjectivity of our responses more effectively than fog. It is the very stuff of hallucination and illusion, as shape-shifting and ambiguous as our own feelings. I once saw a shadow image of myself cast by the low sun against a bank of fog. It was huge and melodramatic. The nineteenth-century Surrey diarist, George Sturt, wrote a note about an early December mist that is fearful and claustrophobic, but also thrillingly portentous, as if, through the vapour, he is sensing the first cogs of the winter solstice beginning to mesh. 'Towards dark, a colourless fog, snow almost gone, and ground soft-oozy underfoot, as though the earth's skin slipped as you trod. A very dark night: no wind; church bells dinning and myself chilly and afraid of the misty evening.'

Nineteenth-century painters were less sensitive to the dark side of fog. It was simply one of their special effects. Moments before he died in 1851, Turner, weather artist extraordinary, was found on his bedroom floor, trying to reach his window to look out at the River Thames. His doctor reported that, just before 9am, 'the sun broke through the cloudy curtain which so long had obscured it splendour, and filled the chamber of death with a glory of light.' That cloud was the pall of soot and sulphur-saturated fog (later christened 'smog') that blanketed London during its nineteenth century hey-day as an industrial city. Painters adored London fog for the way it misted rough-edged city scenes into essays in Impressionism, for its magical softening of detail. It lured Claude Monet over from France. He told his dealer 'what I love more than anything is the fog'. James McNeill Whistler thought the evening smog clothed 'the riverside with poetry', transforming factory chimneys into Italian bell-towers, warehouses into palaces, so that 'the whole city hangs in the heavens, and fairy-land is before us'. Out in the non-fairy landscape, of course, people were dying of asphyxiation, a situation that was allowed to continue until the notorious London smog of December 1952, during which more than 4,000 people are believed to have died as a direct consequence of the fatal combination of soot, sulphur dioxide and cold.

Both visions – the streaked, shifting, glowing fog in Turner's London paintings (Constable called it 'tinted steam') and the suffocating pall out in the streets – are true in their different ways. Weather is a kind of Rorsharch test. We see in it what we need to see, or what we feel is missing from our lives.

And now, each day, we have the inkblot test administered to us virtually, in the shape of the forecast, a ritual which has a significance in our mythology far beyond that of simply predicting the weather. Of course, it's principally a practical tool, made increasingly accurate with the advent of giant computers which can access and analyse second by second changes in pressure and temperature from all over the planet. Manufacturers of ice-cream and umbrellas trust its long term predictions enough to base their seasonal production quotas on it. So do we – at least for a few days in advance – and are willing to use it to make choices about which day to go for ramble at the weekend, or whether to take a raincoat to work tomorrow.

But beyond that, our enthralment with it edges into the realm of magic. The forecast has become an oracle and, like all sooth-sayers, we regard it not just as a source of guidance, but as a scapegoat, a focus for blame when things go wrong. The forecast gives us the opportunity to be as moody as children in front of it. It's an overseeing parent, suggesting how we should behave, recommending spells of gating, reminding us that we are not remotely grown-up or clever enough to have any power or control over the elemental events it is reporting. If its prognostications go badly wrong, we feel we're entitled to throw a tantrum and to blame the forecast (sometimes even the hapless forecaster) rather than the vagaries of the weather itself. When it is right, as it is more often than not, it's no more than we expect of a responsible parent. But even a correct bad forecast still leaves its damp stigmata on the hands of the messenger who delivered it.

So there is a touch of irony in the fact that the very first weather forecasts, which began at a time when superstition held real power, were delivered without our modern magic shows of swooping technicolour pressure fronts and meteorological abracadabra, and in the simplest of ways for the most practical of purposes. I have a copy of *The Shepherd of Banbury's Rules*, first published in 1676, which gives guidance on how to interpret atmospheric signs to foretell imminent weather patterns. The only computing system the good shepherd had access to was an acute eye downloading to a memory bank stretching back over half a lifetime. But it enabled him to make predictions of risk-taking precision: 'A general Mist before the Sun rises, near the full Moon – Fair Weather.' I have no idea whether the Banbury's shepherd's system was statistically valid, but it was based on the same empirical principles as modern forecasting. It never assumed it could help us defeat the weather, but it might enable us to stay one step ahead.

But in 1652, just twenty-four years earlier, the definitive edition of Thomas Hill's hugely popular manual, *The Gardener's Labyrinth*, passed on oracular tips which were precisely about how to turn the weather around. They were based on the ancient principles of sympathetic magic, the idea that like counters like. Firing a gun would disperse thunderstorms. Hanging the ominous pelt of a seal at the entrance of the garden would keep dark clouds away. My favourite is the suggestion that sowing the infamously flatulent seeds of lentil in the vegetable beds would keep them immune from damage by wind.

The progress of forecasting followed pretty much in the pragmatic tradition of the Banbury shepherd. One offshoot was the detailed weather diary of the kind kept by Gilbert White. In 1767, one of his correspondents, Daines Barrington, devised a printed journal proforma, with columns on wind, general weather, plants first in flower and various miscellaneous detail. Its stated purposes were clear, and utilitarian: 'it may also be proper' Barrington advised the journal-keeper, 'to take notice of the common prognostics of the weather from animals, plants or hygroscope … and from many such journals kept in different parts of the kingdom, perhaps the very best and accurate materials for a General Natural History of Great Britain may be in time expected, as well as many profitable improvements and discoveries in agriculture.' Recently, through detailed analysis of these and more recent records it's been discovered that averaging-out descriptions of weather at a given moment of the year is a more accurate predictor of weather now, than the attempt to decipher up-to-the minute reports of the atmosphere's chaotic games.

A crucial difference between these written 'prognostics' and modern forecasts is that they preserved the *archaeology* of the weather. We barely have a weather memory anymore. We imagine lost meteorological fairy-lands and forget the real good times. We turn vague recollections of the routine muddle of poor weather into catastrophic visions of the future. Who now remembers that the summer of 1975 was as long and hot as that of 1976? Snow stopped play at a cricket game in Colchester on 2 June, 1975, but four days later one of the great halcyon summers began. The reason we only remember the summer of 1976 is because, in our doleful weather folk-memory, it was accompanied, *mordanted*, by a drought.

4

HALCYON DAYS

IN MEDITERRANEAN MYTHOLOGY, the king-fisher (*alkuon* in Greek) was believed to incubate its eggs on the surface of the sea, during the spell in November when water and weather were always calm, and which was later known as St Martin's Little Summer. The phrase 'halcyon days' sub-sequently began to be used for any peri-ods of peace and general happiness – and, because these are so often dependent on the weather, for those blue remembered days in which sunshine and bliss are inseparable.

But we shouldn't forget the role of the kingfisher in this, that spark of iridescent azure and cinnamon that is like a flash of fair-weather lightning. In these post-mythological times, of course, kingfishers tend not to raise their young in autum-nal, waterborne nests. But one September morning in the Norfolk Broads, a fledgling perched briefly on our boat as we were having breakfast, just feet from our scrambled eggs. The day that followed wasn't the least bit exceptional in terms of its weather, but it became halcyon because of the benediction of that small flighted rainbow. The kingfisher *stood in* for the sun, becoming a thread in that complex weave of metaphor, ancient association and real physical experience through which we make sense of the weather, and its effect on our feelings.

And because these associations are so personal you can have a halcyon day at any time of the year, and probably in any weather. Coleridge, overjoyed by fatherhood, decreed in the exquisite poetic benediction to his sixteen-month-old son Hartley entitled 'Frost at Midnight', that *every* day should be halcyon to him:

> Therefore all seasons shall be sweet to
>   thee,
> Whether the summer clothes the
>   general earth

With greenness, or the redbreast sit
   and sing
Betwixt the tufts of snow on the bare
   branch …

Even when the weather has been incessantly miserly, you can make a halcyon day from a widow's glimmering mite. The winter of 1979 was notorious for its relentless gloom. On 22 February, I was walking down London's Lower Regent Street, that shadowy chasm of tall buildings, when the sun suddenly peeped through the clouds for the first time in weeks. Quite spontaneously, almost everyone stepped off the pavement into the thin ribbon of watery sunshine in the road, giggling like children in delighted surprise. I'm pretty sure I recall a few brief dance twirls being executed too.

And as the year progresses we all have our personal halcyon moments. On 19 April, 1873, John Ruskin logged an 'Entirely Paradise of a day, cloudless and pure till 5; then East wind a little, but clearing for twilight. Did little but saunter among primroses and work on beach.' I suppose that sums up a climatic Shangri-La most of us would be happy to inhabit. It's the kind of day when we tumble into hyperbole: 'Aah, it was like the First Day of the World!' But one of my spring halcyons, in 2003, was a bit more like the Day *before* the First Day, as if I'd somehow sneaked a glimpse of seasonal evolution still a bit short on its finishing touches.

I need to fill in a little background here. Years ago, when I first began to be fascinated by plants, I came across an obscure scientific conceit, a measure of the advance of spring across the land based on the average first dates on which common wildflowers first come into bloom. Different species flower at times determined by a combination of temperature, rainfall, sunlight and day-length – all the elements that add up to a micro-climate. So, a species – primrose, let's say – that first blooms in mid-April in the sheltered combes of Dorset will be opening simultaneously in the Gulf Stream breezes on the Pembrokeshire coast, and the warm 'urban envelope' of

Kew Gardens. Yet on the exposed scarp of the North Downs, only twenty miles south, it may not be open for another fortnight, and it will be a full three weeks before it graces the windswept hills of Dartmoor and Snowdonia. The lines joining these points of floral coincidence are known as isophenes, and from them it's possible to calculate that spring travels north and east across flat ground at roughly two miles an hour – walking pace in fact, so that it's possible to indulge the fantasy of following it on foot, the guest behind the unrolling carpet.

But spring 2003 had an uncomfortable, out-of-sorts feel about it, at least in the heads of a lot of us. The day before the equinox, the West, always eager to trump nature, invaded Iraq. I felt in need in some kind of seasonal retribution, 'a shot at redemption,' as Paul Simon put it. So the next morning, 21 March, I decided I would walk west from my Suffolk house and meet the spring head-on, as if it might not reach me otherwise.

I guess I was hoping for a straightforward

halcyon, a gorgeous bird of a day, like the fabulous creature which had hatched in the eastern Mediterranean in one of its happier moods. What I found instead was more honestly untidy and indisputably English. The weather itself was typical for March – mild and bright, with a sharp wind blowing from the west, straight in my face. But the isophenes were as tangled as a mad cat's cradle. Celandines and marsh marigolds were in full-bloom on south-facing ditch banks, but barely in bud in frost hollows just yards away. The shady lanes were still busy with winter birds, but when I reached the open sheepwalks of Breckland, there were 'sweeing' lapwings overhead, and the first brimstone and peacock butterflies. I'd sauntered twelve miles west in six hours, so had got no more than half a day deeper into spring. But the micro-landscapes – and their micro-climatic familiars – were so diverse, such a convolution of tumps and dells and thickets, that I'd travelled through about two months of biological weather. It felt a small snub to the violent levelling happening 2,000 miles behind me.

Jan Morris also experienced a political halcyon, a climatic metaphor, in the gilded spring and summer of 1990, which she saw as a fitting farewell to the decade that had given us Chernobyl and monetarism. Nelson Mandela was freed, the Berlin Wall came down, summer birds and prodigious insects swarmed through the blazing sun of May, and on her own day of days, Morris watched seventy-seven pipistrelle bats fly out of their roost above the kitchen of her Welsh cottage. Those days she wrote, were 'an allegorical moment of reconciliation ... through which all too briefly flickered a message that the worst might be over.'

Of course it wasn't over, nor did 1990 presage a run of New Age weather. As so often, we British have had to continue scratching for halcyon moments in improbable situations, often by being willing to laugh at our fantasies and the all too frequent ghastliness of our climatic lot.

Sometimes such moments are the stuff of pure farce. On 3 July, 1996, play in the men's quarter-finals at Wimbledon, epicentre of the English summer, had to be abandoned for almost the entire day because of torrential rain. Cliff Richard, well-known as a tennis nut, happened to be in the covered part of the Centre Court, and the organisers, liberated from their usual reserve by desperation, had the wheeze of asking him to sing for the increasingly bored and bedraggled crowd. There followed two hours of surreal vaudeville, with the computerised scoreboard printing out the lyrics as Cliff belted out his hits, accompanied by a scratch WLTA backing group led by Martina Navratilova, and a Centre Court crowd by then in full carnival mood.

We should never forget that the halcyon is a water bird.

My own best halcyon day, perhaps one of the most idyllic of my life, was also involved with water, but had a twist in the tail. It was the mid 1970s, and I was in the north-west of Scotland with the late, great photographer Tony Evans, looking for the alpine wildflowers we had so dismally failed to locate up in the mountains. The alchemy of the isophenes sometimes brings

them down to sea-level and we struck lucky
by the shores of Loch Linnhe – tight tufts
of yellow saxifrage and drifts of succulent-
leaved roseroot ringed the great lagoon, and
we sprawled out among them. It was a day
of the purest Highland light, no wind, the
sky an almost opalescent blue. Tony set up
his camera on a bank about fifty feet above
the loch and I went down to the edge, to lay
a bottle of white wine to cool in the water,
marking its position, if I remember right,
with a sock. We lounged there most of the
day. I held an elegant white parasol over
the camera to shield Tony's fragile gelatine
filters from the spray of a waterfall close
to our chosen saxifrage. A seal, barking
quietly, swam up the loch, and, one hour
later, swam back again. By mid-afternoon
it was almost too hot to work and we were
relieved when we finally got the shot in the
bag. Never has the prospect of a chilled
Chablis seemed so ambrosial, and I went
down to the loch-edge to retrieve the bottle, only to find it – and the sock marker –
had vanished. I'd failed to take into account
the fact that Loch Linnhe was tidal and

that our now urgently-needed refreshment
was, in essence, fifty metres out to sea. We
spotted it eventually with our binoculars,
glinting mockingly deep in the water and,
as wine-steward for the day, I had the duty
of retrieving it, a task which involved a
breath-stopping dive under water twenty-
five degrees cooler than the bank of wild
thyme we had been basking on.

Of course, it tasted all the more necta-
rous for the wait and the effort, and was a
reminder that – except for the rare occa-
sions when they descend on you unbid-
den – days of halcyon weather often occur
because in some way, often unconsciously,
we have worked for them. Our journalis-
tic leading-man, Gilbert White, had many
small-scale halcyons, for instance on hot
July mornings when his beloved swifts,
'getting together in little parties, dash
round the steeples and churches, squeak-
ing as they go in a very clamorous manner.' But his weather epiphany occurred
one autumn, on 12 September, 1758, when
he was able to hold a melon feast on the
steep beech-clad hill behind his house, the

culmination of years of attempting to grow this sub-tropical fruit in the fickle English climate. Eighteenth-century naturalists and gardeners were infatuated by melons. More than any other vegetable growth they seemed to embody the ideals of the Age of Enlightenment. They were exotic, picturesque, and repaid investment and scientific ingenuity with enormous productivity.

Throughout the 1750s the 'melon ground' as White grandly called it, was the epicentre of activity in his garden. And each year, as he nursed these temperamental fruits to maturity, he became locked into his own version of our common struggle with the rigours and vagaries of British weather. His melons hung, in late winter, in a precarious balance between succumbing to damp-induced mildew or freezing to death; and later, between suffering drought or sunburn or being flooded out. *Plus ça change*. Their fortunes, and his reactions, are, as always, meticulously recorded in his journal. On 21 March, 1758, there was heavy snow, and a 'stinking, wet fog ... Very trying weather for the Hot beds'. On

16 April: 'So fierce a frost with a South-wind as to freeze the steam which run out in water between the panes of ye Melon-frames into long icicles'. Two months later it was too hot, and the melon leaves were 'strangely blistered' by being in fierce sun while the dew was still on them.

But in mid-August, at last, it all comes good. Gilbert cuts the first 'Cantaleupe', and finds it 'perfectly delicate, dry, & firm [despite] the unfavourable weather ever since the time of setting.' Then, on 12 September, with all the *homage* you might show to a classic Bordeaux vintage, he holds a ceremonial melon feast in his little hillside hermitage, and cuts up a brace and a half of fruit among his fourteen guests. 'The weather', he adds 'very fine ever since the ninth.'

But there can be winter halcyons, too, days of clarity and acute perception quite unlike the luxuriance of the warm months. I'm not a winter person myself. I can marvel at the sight of a landscape made anew, reduced to its fundamentals by snow. But I can't stop thinking about what may be happening underneath its virginal finery. What is starving? What has already died? I once saw thousands of migrating red-wings blown exhausted by a blizzard onto the Norfolk coast, their emaciated bodies already like dark absences against the whiteness. But the Romantics discovered something extra in winter, the possibility of *accelerated solitude*. Winter offered an unexpected *monde renversé* for those who cared to look beyond its snuggled interiors: the chance for fierce physical engagement with nature, but on your own, not picnicking with your peers in the Melon Ground.

And ice-skating, brought to Britain by the Dutch in the mid-seventeenth century was the obvious way in. Gilbert White and Francis Kilvert were mad for skating. So was Goethe, and there is a deliciously wry painting of him from the 1850s, looking a little like Lord Byron and swooping across the ice through crowds of doting female admirers. But it was Wordsworth, legendary as a Lake District skater, who most perfectly captured the Romantic electricity of the solitary ice-glider, the spark and hiss of the frost-fisher. There is a long passage celebrating the jollity of communal night skating in Lakeland in his masterly poem, *The Prelude*. But halfway through he swerves sideways, 'leaving the tumultuous throng,/ To cut across the reflex of a star/ That fled, and, flying before me, gleamed/ Upon the glassy plain ...'

In the hard winter of December 2011, the ponds on our village green froze solid. I don't skate, but my partner Polly does. She was brought up in the Norfolk Broads where her father was a country GP, and a renowned skimmer of the lakes and fens. She has inherited his skill and guts – and his skates. So on Boxing Day, we walked down to the biggest of the horse ponds, which, in width and length, is about the size of a cricket pitch. Poll put on her on her own skates, and after a little initial unsteadiness, began to glide demurely about. I felt I should keep a weather eye on her, so I meandered round the edge of the common, casually looking for field-fares and barn owls, and enjoying the way the icy crust over the mud scrunched like a crème brûlée under my feet. Then I spotted a tall dark stranger walking briskly in our direction and felt I should head back to the pond. He proved to be an amiable Dutchman in his mid-sixties, playing truant from his mother-in-law. He looked enviously at Polly's twirling, and she asked if he'd like a go. Being a Dutchman, his answer was obvious. What was even more eerily coincidental than his serendipitous arrival was that his skate size was the same as Polly's father's, whose sleek sixty-year-old leather-shoed blades – so different from the miniature canoes Dutch skaters wear – she had generously brought along for just such an eventuality. Within seconds our new friend had wriggled into them, and was away over the ice. He needed a while to adjust to the skates' unfamiliar slimness, but was soon arcing round the little pond in true fenland – and true canal-land – style, fast but languorous, one hand behind the back, the other sweeping like an elegant pendulum in front. As he speeded up, the ice began to hiss, and blow in clouds as he took the corners with beautiful, double-pace foot-crossing. As the sun set he became almost a blur against the frosted scrub, and I thought he was one of the most sublime cold-weather creatures I had ever seen. He made a halcyon day for me. But, skimming across the glassy element on which he must surely have been hatched, he brought this tale full circle and became a winter halcyon himself.

5

MIXING MEMORY
AND DESIRE

THE 27TH OF FEBRUARY 2019 has been officially proclaimed the hottest February day 'since records began'. At least that was the diplomatic description. It may well have been the hottest since the Ice Age. At Kew Gardens the thermometer reached 21.2 C, and the mean maximum temperature for the month was 3.5 C above the average for the last thirty years. But to tell the truth, beyond the unprecedented temperature, it was much like any fine day in late winter. Bumble bees and brimstone butterflies were flying, the cherry plums were in blossom, and the wild daffodils on our meadow, the very same species that Dorothy Wordsworth observed as they 'reeled and danced' on the shore of Lake Ullswater, and which inspired her brother's famous poem, were poised to be in wavering bloom at their usual mid-March date. And everything was bathed in that special new year's light, so different from the rich glow of fine days in late October. This early in the year there are no leaves on the trees to interrupt the sun, and the landscape seems dusted with the faintest hint of pollen.

So we donned T-shirts, ate out of doors, went for familiar walks that had never looked just so. But as an *experience* it was just another perfect early spring day, and by next year we will have forgotten it. Spells of weather are analogous to human moods in many ways, not least in the way we use one as a source of metaphors for the other. But unless they create something indelible – a breakdown, a house destroyed – they become hazy and amorphous in recollection.

The press was unreserved in its delight. 'FABruary' ran the *Sun*'s front page headline. But in the margins there were more ominous bulletins. Meteorologists were dumbstruck by a temperature whose probability they said was close to zero. Swallows appeared, recklessly, in Dorset. On

Saddleworth Moor in Lancashire, scene of raging heather blazes in the heat of the previous summer, wildfires returned. There was a high chance that the premature spring we were relishing was yet another 'extreme weather event', a collateral effect of global warming. Will we forget that possibility, just as we will forget the unseasonable ice creams?

T. S. Eliot's poem *The Waste Land* famously begins with the line 'April is the cruellest month'. This isn't a reflection on washed out bank holidays. In Eliot's metaphor April stands for spring, for new possibilities, sudden spurts of growth, calls to action, and he is reminding us that expectation and hope can be disruptive emotions, not just because they may be unfulfilled, but because they may disturb our comfortable torpor. April 'breeds lilacs out of the dead land' forces us to *feel* again after our winter somnolence. April mixes 'memory and desire'. It may not be a pleasant experience, being woken up like this. We would have been less than human not to have relished the halcyon days of February 2019, but shouldn't it also have shaken some of our complacency about climate change?

But weather and memory have never had a good fit in our psyches. It's as if, in a climate of constant uncertainty, we cannot bear the risk of recalling it with exactitude, for fear the memories might either strike us down or fill us with hopeless expectations. So we invent mythical Golden Ages, and forget the real good times. False memories of bad weather become our default expectation. The weather of the last week of February in the previous year, 2018, has been elevated into a mythic winter cataclysm simply because someone euphoniously dubbed it 'the Beast from the East'. In fact, it was an unexceptional winter spell that lasted just eight days and caused no more than average disruption. Most people remember the great storm of October 1987 not for the devastation it caused, but for Michael Fish's gaffe on the televised forecast, guaranteeing no hurricane was on the way.

Over the years I've keep a sporadic and impressionistic diary of weather and natural events, and reading it back, I'm always shocked at how little of the details I can recall without this prompt. I remember the 'events' but they are dislocated, out of order, not attached to any particular year. I thought I would call up the entries for February 2009, ten years before, and on the surface, an entirely unmemorable winter month. But it had a special quality, which I had quite forgotten.

My journal records the snow beginning on 2 February, a Sunday. Polly and I, and her two Norfolk grandchildren, had arranged to meet Robert Macfarlane and his daughter and two others (all under five) at Horsey Mere in the Norfolk Broads:

A clear day, but piercing East wind, blowing brief horizontal blizzards … The kids are heroic. We shelter under one of the big oaks along Waxham Dyke during a white-out. The children sit in a row – Babes in the Marsh – and Polly begins to thatch them with reeds from the dyke, weaving them under their head scarves and draping over their shoulders and chests. They yelp with a mixture of frost-pinch and delight, and Rob – ever the wordsmith – defines a squall as 'a flurry of snow and the noise you make when it hits you' … We need to move on, and the kids, now in fits of giggles, hold onto their thatched roofs. They look like mobile woodlet from a Shakespearean comedy.

By the end of the week the country was in chaos:

London brought to a total standstill. Salt running out everywhere, and some local authorities are buying in table salt … Both Severn bridges were closed because great slabs of ice were crashing through car windows from overhead road signs …

And our high jinks on the marsh were being echoed on a national scale:

The people play truant! Most schools

in East Anglia have been shut down for at least one day, and many of the younger kids have never seen snow like it. They are improvising toboggans out of old carpet and strips of plasterwork from skips ... In Swindon, fifty ten-year-olds are suspended from school after refusing to come indoors after morning break. But in Chipping Norton, the kids build an igloo on their playing field – and spend the night in it ... There are 15,000 complaints to the emergency services about snowballing! ... We are informed by the government that all this skiving off has cost the nation £1.2 billion.

I have vivid recollections of our day on the marsh, but not the year it occurred, or that it happened in the very same cold snap as this joyous and anarchic breakout by the country's schoolkids, engaging with the weather in a way their parents too often avoid.

Yet if our weather recollections are fragile and unreliable, the landscape has its
own memory bank. I had moved to East
Anglia six years before this ludic winter,
in the autumn of 2003, and what followed
was a carnival season of rough weather.
Ice, flood and wind followed each other in
quick and successive rotation. The building I was lodging in was a sixteenth century
half-timbered farmhouse, and as draughty
as a barn. When a freezing gale hit us in
November, it burst through the beam
joints and window cracks. It held the cat
flap open in a horizontal position, and was
palpable deep inside the house. I stuffed
old pillows up the chimney and rolled-up
bin liners in the biggest knot-holes in the
floor, but could still feel it rasping my face
in bed. The house was hyperventilating,
gasping in cold air, and exhaling any internal warmth through every orifice.

Yet something else was circulating
through this accidental air-conditioning.
A strange miasma began to drift into the
rooms through the joints and knot-holes, an
aromatic, airborne flotsam of rotten wood

crumbs, lime-plaster dust, wisps of horse-hair from the ancient wattle-and-daub, and old swift droppings, sucked up from the loft where they nested and whirled down through the cracks in the ceilings. This was an aerial time capsule from the seventeenth century, a fossil animated by the wind, and evidence that weather, seemingly so much a phenomenon of the *now*, has currents reaching back into the past and forward to the future. The first intimations of the man-made global warming that was to follow have been discovered deep in the Antarctic ice. Layers dating from 10,000 years ago, the time when farming and therefore civilisation began, show a small but definite spike in trapped carbon dioxide.

Gilbert White, this book's weather laureate, presents evidence of his village's past weather which is as vivid as the strata in an archaeological dig. Selborne in Hampshire has a network of sunken lanes, worn deep into the sandstone rocks by traffic and weather. They're more than just a system of byways, the community's internal and external connective tissues. They are landmarks, physical records of the past history and everyday experience of the parish. Every extreme of weather – ice, gale, landslide – leaves lingering traces here. In the eighteenth century they became blocked during snowfalls, and flooded after heavy rain, to the extent that Selborne was often cut off from the outside world. It may have been this isolation which prompted White to become an obsessive writer of letters, the most brilliant of which form the core of his revolutionary book *The Natural History of Selborne*. The hollow lanes were active forces in his life, habitats which prompted him to reflections on history and ecology, and occasionally, whimsy. This is how he described them in one of those letters, addressed to the naturalist Thomas Pennant (though probably never sent):

These roads, running through the malm lands, are, by the traffic of ages, and the fretting of water, worn down through the first stratum of our freestone, and partly through the second; so that they look more like watercourses than roads... In many places they are reduced sixteen or eighteen feet beneath the level of the field; and after floods, and in frosts, exhibit very grotesque and wild appearances, from the tangled roots that are twisted among the strata, and from the torrents rushing down their broken sides; and especially when those cascades are frozen into icicles, hanging in all the fanciful shapes of frost-work. These rugged gloomy scenes affright the ladies when they peep down into them from the paths above, and make timid horsemen shudder while they ride along them; but delight the naturalist with their various botany, and particularly with their curious *filices* [ferns] with which they abound.

I've walked Selborne's hollow lanes many times (once on an early March day very like our February 2019 prodigy) and their rough-hewn evidence of the village's eventful climatic past is still unmistakeable. So is their 'various botany', in the precise addresses White gives for it. In the middle

of a fierce November frost he found poly-
pody fern and gladdon iris 'in the hollow
lane between Norton-yard & French-meer
just without the gate'. The iris, he remarks,
'was thrown in all probability out of the garden which was formerly on the other side of
the Hedge'. Wall lettuce, male and hart's-
tongue ferns were 'in a most shady part of
the hollow lane under the cover of the rock
as you first enter the lane in great plenty, on
the right hand before you come to nine-acre-
lane.' They can still be found in these same
spots today, active botanical memories.

The plants which hold the most expressive weather memories are trees. Each year
the trunk of a tree puts on an annual ring,
a cylinder of new wood surrounding the
existing core. The rings are clearly visible,
and measurable, when a trunk is cut hori-
zontally across, and are the chief contribu-
tor to the 'grain' of wood. Counting them
gives the age of the tree in years. But each
one is different in breadth, texture and colour. Wet summers tend to generate more
soft woody growth, and thus broad rings;
drought years to produce narrow, denser

annular wood. In close-up it is sometimes possible to see the influence of a particular season's weather inside the annual ring. In trees more than fifty years old, the very narrow ring produced during the drought year of 1976 is striking.

The study of tree ring growth is called dendrochronology, and it has produced a fascinating index for dating ancient wood. In any species of tree in a particular weather region, the pattern of the annual rings is very similar. This means that the unique sequence of annual rings – a kind of arboreal bar code – can be used as a key, against which wood of unknown age can be dated. Older wood will have a ring sequence in which only the youngest parts correspond to this key, meaning the older sequence can extend the master key further back in time. It's sometimes possible to measure the annual rings on bog oak and other fossil trees; and combining this with dating the tree remains by calculating their radioactive carbon traces, means that, in some areas of Europe, it has been possible to construct an almost continuous key going back into prehistory. This also provides a rough guide to the weather in the growing seasons of any year. The annular sequence persists in wood which has been crafted, so, if the tree species and area of growth are known, the key can be used to help date pieces of furniture, and reveal the weather narrative their raw material grew through.

In French, the word *temps* means both time and weather, suggesting all kinds of interesting linkages between the two. Eco-systems evolve in time precisely because of the influence of weather and climate. The history of weather is locked into the wooden cases of clocks and patterns of bird migration, into the formation of glaciers, and their melting. We experience weather overwhelmingly in the present. But it is founded on complex patterns stretching far into the past, which have momentous implications for the future. Trying to resist our habitual weather amnesia might help us think more urgently about what is to come.

6

THE STORM
CLOUDS OF THE
TWENTY-FIRST CENTURY

'SUNSHINE IS DELICIOUS, rain is refreshing, wind braces us up.' So chirped that arbiter of Victorian cultural values, John Ruskin, ending his eulogy to climatic variety with one of the most quoted of all weather sayings: 'there is really no such thing as bad weather, only different kinds of good weather'. That rather neatly sums up the idea I've been exploring in this book, that weather, an incontestable feature of the physical world, is also a creature of our imaginations. How we experience and deal with it depends on our moods and memories and powers of myth-making, on how we talk to each other, on our hopes for the future.

For Ruskin the full truth of this became horribly clear years later, when his normally sharp and insightful journal becomes increasingly infused with dark visions.

From the spring of 1871 he became convinced he could see an immense storm-
cloud louring over the Lake District, often
accompanied by 'plague winds of a diabolic
aspect'. On 17 July he records an evening
that had become 'the blackest ... the devil
has yet brought on us – utterly hellish,
and the worse for its dead quiet – no thunder or any natural character of a storm ...
the black *shaking* was worst of all.' When
thunderstorms did materialise they were
like 'railway luggage trains ... the air one
loathsome mass of sultry and foul fog like
smoke.' Soon he was beginning to believe
the ferocity in the sky was directed against
himself. He was losing his ability to draw
natural forms and, in 1880, after recording
'wild wind and black sky – scudding rain and
roar – the climate of Patagonia instead of
England,' he confessed that he was in a state
of 'hopelessness, wonder and disgust ... as
if it was no use fighting for a world any more
in which there could be no sunrise.' Four
years later, in February 1884, he delivered
two lectures on his observations at the London Institution, entitled 'The Storm-Cloud

of the Nineteenth Century'. He summed up the new weather in a phrase that has a chilling resonance today: 'Blanched sun, blighted grass, blinded man.'

Ruskin's mental health had been deteriorating since the 1860s, and there's no doubt that these terrible visions were a consequence of clinical depression and maybe episodes of true paranoia. But they may also have been partly real. The blast furnaces at nearby Barrow and Millom were increasingly polluting the air with soot and fumes and the invisible effluvium of carbon dioxide – the storm cloud of our own times.

Yet dismissals of Ruskin's visions as either psychotic hallucinations or simple melodramatic exaggerations of real industrial-age weather seem inadequate. For all their mania, they chime unsettlingly with our current fears and social anxieties. At their heart was a frightened man ill-at-ease with himself, with the blind and insensitive advance of technology (that railway luggage train!), with nature's seeming indifference to humankind, and our indifference to nature. A man projecting onto the weather his feelings about the state of the world. Someone rather like ourselves, in fact, as we gaze nervously at the gathering storm clouds of the twenty-first century.

Twenty-twelve was declared the wettest year in England for at least a century. The flooding, especially in the Midlands and West Country was devastating. During November and December 8,000 homes were submerged and the Environment Agency issued 1,000 flood warnings. The whole of Devon and Cornwall were cut off from the rest of England for days, first by rail and then by road. Up here in the flatlands of East Anglia, where flood-water generally disperses itself widely but thinly across the whole landscape, we were living in a rural time-machine, the ancient dips and subtle gradations of the land pricked out by water just as they must have been at the melting of the glaciers 15,000 years ago.

It's hard to believe that only nine months ago we were in the middle of a drought that the water industry informed us would take two winters of heavy rain to cancel out.

The spring was so cold that bees bunkered down in their hives in hibernation mode. The apples weren't fertilised, resulting in the worst crop for fifteen years. The summer lasted about a week. Then, in autumn, the increasingly unstable jet-stream flipped again and the monsoon rains began. Hose-pipe bans were rapidly converted into flood alerts. Potato crops were simply washed out of the ground. Ash die-back, a new tree disease in Britain, was rumoured to have migrated here from the Near East on warming air contours. In November, the popular press, gleefully tapping into the growing sense of climatic apocalypse, predicted December would begin with the worst winter freeze-up for 400 years, which was followed, it hardly needs saying, by one of the warmest, if wettest, of winter months. It seems as if the whole pattern of seasonal weather, as well as our capacity to talk sensibly about it, has gone completely out of kilter.

These oscillations, taking Britain's always unstable weather to a new state of *reductio ad absurdum*, have been going on long enough to qualify as a trend. Are they also the consequence of global warming? Quite probably. Scientists have long suggested that in our corner of the Atlantic the results of climate change won't be a pleasantly gentle rise in temperature, making England into a kind of northern outpost of the Dordogne, but swings between extremes of weather, with droughts, heavy rain and strong winds likely to be the dominating features. In other words, the traditional British mixture as before, only worse and more muddled.

This book hasn't been about climate, or its changes. It's been a personal look at how we live with the weather that is our daily, intimately experienced embodiment of climate. But if the climate itself is on the move, then it becomes part of this story.

Only those with ideological blinkers or vested interests deny global warming is happening, and that human activity has a major role in it. But I wonder if a similar kind of denial, a refusal to accept extremely uncomfortable likelihoods, is blinkering those who believe we may be able to halt it.

My own view, if I may be forgiven one last meteorological metaphor, is that we have a snowball's chance in hell of stopping it, at least in the short term. The last twenty years have seen nothing but missed targets and repeatedly postponed agreements. Politicians are too self-interested, corporate business too greedy, scientists barely able to grasp the complexity of what is happening, and the rest of us, the buck-passing public, too irrevocably wedded to our high-consumption lifestyles. That doesn't mean we should stop trying. It would be good to think we were mature enough as a species to pull this off; yet I wonder if we could tolerate the authoritarian governance and high-risk planetary engineering which would be necessary even if we were to find a solution.

There have been utopian schemes for improving the climate for centuries, and they occupy one of the more bizarre – and worrying – corners of our weather mythology. The eighteenth-century historian Edward Gibbon believed the earth's climate was improving, and would continue to do so, because of man's vigorous taming of nature, especially the clearance of forests, 'which intercepted from the earth the rays of the sun'. He *yearned* for global warming. Schemes put forward by the engineer Hermann Segel in the imperious mood of Germany in the 1920s and 30s involved draining the Mediterranean and flooding the Sahara desert. That this would wipe the Belgian Congo off the map was regarded as a minor technical difficulty. As late as the 1960s, the Russian engineer P. M. Borisov put forward a plan to dam the Bering Straits and divert the warm water of the Pacific into the Arctic Sea, thus melting the ice-cap completely and making Siberia warm enough to grow cherries.

I find it hard to see much difference between these mad fantasies, with all their arrogance and ecological ignorance, and current schemes to ameliorate global warming with big technological fixes. Sowing the oceans with thousands of tonnes of soluble iron, for instance, to promote the growth of plankton, which would mop up atmospheric carbon dioxide. Or sending giant parasols into space, to shield us from the sun. The unknown and unpredictable ecological effects that could result might be as catastrophic as the warming they are intended to prevent. All these wild dreams reflect our seemingly unshakeable belief that we are clever enough to control nature – the same hubris that got us into the climate crisis in the first place.

Grandiose schemes have no part in the way earth – the intricate, diverse, locally inventive earth – works. So my forlorn guess is that we will have to confront climate change in the way the planet has always done, muddling and adapting our way through as best we can. It's likely to be a rough ride. More human populations will starve. Many wild species may become extinct. Some landscapes will change in ways we can't imagine. But what we do not know is exactly where or how all this will happen. We may be able to make statistical generalisations about climate shift, but not about the complex weather it will generate, or about how we and all other living things will react. Succumbing to

Ruskin's doom-laden storm-cloud will get us nowhere.

I've experienced two major climatic crises in my life, and seen something of the ways humans and other beings respond. The first was short and dramatic, the great storm of 16 October, 1987. It lasted just five hours and toppled fifteen million trees, less than one per cent of south-east England's total. But it was a cultural apocalypse, which changed the way we think about the apparent stability of nature, and about the kind of relationship we should have with it. It revealed our touching affection for our arboreal neighbours, and a dismal lack of understanding about how they themselves lived as a community. I roamed about the shattered woodlands for a whole week after the storm and saw and heard stranger images emanating from the human observers than from the tumult of tumbled, creaking, odoriferous, splitting timber itself. Foreboding, guilt, anger at what were thought to be random malign forces, were as clamorous as the ubiquitous whine of chain-saws. At Kew they held an informal service for 'the fallen'. The Tree Council issued a press statement, an extraordinary solecism that seemed to place the republic of trees solidly inside the kingdom of man: 'Trees', it solemnly warned, 'are at great danger from nature'. Some heritage pundit went on television and declared that 'the landscapes of southern England had been ruined for ever'.

Next spring things looked rather different. Except where landowners had sent in the bulldozers to clear away the wreckage (and most of the underlying soil) the newly sunlit woods were covered with seedling trees, as thick as grass in places. Ten years on and they had grown into woodlets twenty feet tall. Now, a quarter of a century later, it is hard to tell where the storm hit, and the idea that catastrophes are an entirely natural and often renewing phase in the evolution of woodlands has become part of conventional wisdom.

My second experience of climate change is long term and ongoing. Ten years ago I moved from the storm-tossed woods of the Chilterns to the wetlands of Norfolk, where

flooding is the major threat. Not flooding in the manner of the sudden torrents
in the steep river valleys of the West, but
an insidious rising of the waters under the
earth, slow seepages in from the sea. Much
of East Anglia lies only metres above the
sea-level and the whole geological region is
slowly tilting down towards what is locally
still called the German Ocean. The warming sea is, for its part, slowly rising, and
combined with increasingly frequent storm
surges, is causing breaches of the sea-walls
on an almost annual basis. The villages on
seaward edge of the Broads are slowly tumbling into the sea, and the graves of some
of Polly's ancestors will one day be as deep
in the North Sea as Dunwich church. Further north, attempts to keep the sea out with
ever-higher shingle banks has simply given
the waves bigger targets to ram-raid. After
the great storm surge of 9 November, 2007,
they looked like chewed fish. The sea had
gnawed through in dozens of places and
sprayed tongues of shingle hundreds of
metres into the freshwater marshes where
bittern and marsh harrier breed.

There is nothing to be done about this. The bill for building one metre of sea-wall reliable enough to keep the sea out for a maximum of twenty years is £10,000 – which would amount to £20 billion for the entire eastern coastline between Ramsgate and Hull. Imagine the feelings of inland taxpayers confronted with that bill. Instead the official policy is one of 'managed retreat', defending settlements and vital ecosystems wherever feasible, but allowing the sea in elsewhere, to form new, natural saltmarsh, the best absorbent buffer against tidal surges.

To some, this smacks more of reckless surrender than managed retreat. But this is where the narrative of East Anglia's vernacular engagement with the weather diverges from that of much of Britain. Indundation by the sea has been a constant in East Anglia for millennia, and it is hard-wired into the indigenous folk-memory. There has been an accommodation reached that hasn't always been possible in areas where weather extremes are less predict-able. If there's been a long struggle to keep the water out, there's also been an irresistible temptation to invite it in, to the imagination and the heart. East Anglians are conjurors with water. They eulogise it, paint it, coax it into extraordinary forms and structures, and generally treat it with the same respectful chutzpah that a snake charmer shows to a snake. Even during the terrible storm surge of 31 January, 1953, in which 307 people lost their lives, this spirit survived. In the bar of the Jolly Sailors in the Suffolk coastal village of Orford, there's a brass plaque on the wall marking the level reached by the floodwaters. The locals kept drinking by taking to the table-tops, while the landlord dived down heroically into the cellar to bring up new barrels of beer.

We have a boat on the Norfolk Broads and spend a lot of time in what is likely to be the front-line of tidal flooding as the warming sea rises. The whole area may become saltmarsh and human communities and freshwater ecosystems will have to be given room to migrate westwards. The irony is that the Broads were themselves the creations of climate change. They began as medieval open-cast peat mines, and were then flooded by an unexpected rise in sea-level and decades of torrential rain in the thirteenth century. But the locals adapted to their new swampland habitat. They designed special shoes, called pawts, for walking in marshland. They built houses on stilts.

Five hundred years later, with the Broads now a National Park, and one of the most spectacular wild landscapes in Britain, that ancient accommodation continues. Riverside villages are full of self-build wooden cabins. The first stilt-houses are reappearing. Solar-powered boats glide about the water, and already household rubbish is collected by bin-barges. Our national uncertainty about what the weather is going to do often prevents us from making this kind of commitment. We build the wrong kind of houses in flood prone areas, or no houses at all. Why not create buildings to *coexist* with water? Venice and Amsterdam seem to have made quite a good job of it. It could be a first step in learning to live *with* nature and climate change, as well as doing our feeble best to slow it down. Coming home in our boat on autumn evenings we sometimes see a strange and beautiful weather metaphor for this, a *harg*, a sea-mist, that constant symbol of ambivalence, which blows back and forth across the coastal boundary between ocean and broad, so that churches and windmills are continuously appearing and disappearing and we scarcely know if we are at sea or inland.

Meanwhile we will doubtless continue with our tragicomic street theatre of daily coping. Parishioners will rope themselves to favourite trees to try and keep them upright in gales. Policemen will improvise giant snowballs to stop off slip-roads on iced-up motorways. Crowds at sporting events will sing to frighten away increasingly torrential downpours. And all the while, waving *and* drowning, we will say to each other 'It's turned out nice again.'